P9-BXY-536

3 5674 02141239 1

J621.042 R96e

DETROIT PUBLIC LIBRARY

LINCOLN BRANCH LIBRARY
1221 E. SEVEN MILE
DETROIT, MI 48203

DATE DUE

JUL 3 1 1995

JUN 0 5 2000

BC-3

2/00

LI

FEB   1995

LINCOLN BRANCH LIBRARY
1221 E. SEVEN MILE
DETROIT, MI 48203

■ Science Experiments for Young People ■

Environmental Experiments

About

RENEWABLE ENERGY

Thomas R. Rybolt

and

Robert C. Mebane

ENSLOW PUBLISHERS, INC.

Bloy St. and Ramsey Ave.   P.O. Box 38
Box 777   Aldershot
Hillside, N.J. 07205   Hants GU12 6BP
U.S.A.   U.K.

j 621.042
RYB

A journey is easier with a light to follow. I had three of the best and brightest to guide my way. For my sisters, Alice and Arlen, and my brother, Bill—TR

For Carmen—RM

Acknowledgments
We wish to thank Mickey Sarquis, Ron Perkins, and Clarence Murphy for their helpful comments on the manuscript.

Copyright ©1994 by Thomas R. Rybolt and Robert C. Mebane

All rights reserved.

No part of this book may be reproduced by any means without the written permission of the publisher.

Library of Congress Cataloging-in-Publication Data

Rybolt, Thomas R.
    Environmental experiments about renewable energy / Thomas R. Rybolt and Robert C. Mebane.
        p. cm. — (Science experiments for young people)
    Includes index.
    ISBN 0-89490-579-1
    1. Renewable energy sources—Experiments—Juvenile literature. [1. Renewable energy sources—Experiments. 2. Experiments.] I. Mebane, Robert C. II. Title. III. Title: Renewable energy. IV. Series: Rybolt, Thomas R. Science experiments for young people.
TJ808.2.R93 1994
621.042′078—dc20   93-48543
                                                      CIP
                                                      AC
Printed in the United States of America

10 9 8 7 6 5 4 3 2 1

Illustration Credit: Kimberly Austin

Cover Illustration: © Spencer Swanger / Tom Stack and Associates

LINCOLN BRANCH LIBRARY
1221 E. SEVEN MILE
DETROIT, MI   48203

CONTENTS

Introduction

Earth

Earth, our home in space, has supported life for billions of years. But with a growing human population, people are having a greater effect on the environment than ever before. Together we must learn about the problems facing our environment and work to protect the earth.

There are many ways we can work together to protect the earth. We can ask adults to use more fuel-efficient cars (cars that get more miles per gallon of gasoline). We can ride bikes or walk instead of getting rides in cars. We can recycle aluminum, paper, plastic, and glass, and we can plant trees. We can save energy by turning off lights when they are not in use. We can save energy by not keeping rooms and buildings too hot in the winter or too cold in the summer. Another way we can help the earth is to learn more about the environment.

This series of environmental books is designed to help you better understand our environment by doing experiments with air, water, land, energy, and life. Each book is divided into chapters on topics of environmental concern or importance. There is a brief introduction to each chapter followed by a group of experiments related to the chapter topic. This series of environmental experiment books is intended to be used and not just read. It is your guide to doing, observing, and thinking about your environment.

By understanding our environment, we can learn to protect the earth and to use our natural resources wisely for generations to come.

Atoms and Molecules

Understanding something about atoms and molecules will help you understand our environment. Everything in the world around us is made of atoms and molecules. Atoms are the basic building blocks of all things. There are about 100 different kinds of atoms. Molecules are combinations of tightly bound atoms. For example, a water molecule is a combination of two hydrogen atoms and one oxygen atom.

Molecules that are made of only a few atoms are very small. Just one drop of water contains about two million quadrillion (2,000,000,000,000,000,000,000) water molecules.

Polymers are large molecules that may contain millions of atoms. Important natural polymers include natural rubber, starch, and DNA. Some important artificial polymers are nylon, which is used in making fabrics; polyethylene, which is used in making plastic bags and plastic bottles; and polystyrene, which is used in making styrofoam cups and insulation.

Atoms are made of smaller particles called electrons, protons, and neutrons. The nucleus is the center of the atom and contains protons and neutrons. Protons are positively charged, and neutrons have no charge. Electrons are negatively charged and surround the nucleus and give the atom its size.

Atoms and molecules that are charged are called ions. Ions have either a positive charge or a negative charge. Positive ions have more protons than electrons. Negative ions have more electrons than protons. Sodium chloride, which is the chemical name for table salt, is made of positive sodium ions and negative chlorine ions.

Atoms, ions, and molecules can combine in chemical reactions to make new substances. Chemical reactions can change one substance into another or break one substance down into smaller parts of molecules, atoms, or ions.

Science and Experiments

One way to learn more about the environment and science is to do experiments. Science experiments provide a way of asking questions and finding answers. The results that come from experiments and observations increase our knowledge and improve our understanding of the world around us.

Science will never have all the answers because there are always new questions to ask. However, science is the most important way we gather new knowledge about our world.

This series of environmental experiment books is a collection of experiments that you can do at home or at school. As you read about science and do experiments, you will learn more about our planet and its environment.

Not every experiment you do will work the way you expect every time. Something may be different in the experiment when you do it. Repeat the experiment if it gives an unexpected result and think about what may be different.

Not all of the experiments in this book give immediate results. Some experiments in this book will take time to see observable results. Some of the experiments in this book may take a shorter time than that suggested in the experiment. Some experiments may take a longer time than suggested.

Each experiment is divided into five parts: (1) materials, (2) procedure, (3) observations, (4) discussion, and (5) other things to try. The materials are what you need to do the experiment. The procedure is what you do. The observations are what you see. The discussion explains what your observations tell you about the environment. The other things to try are additional questions and experiments.

Safety Note

Make Sure You:

- Obtain an adult's permission before you do these experiments and activities.
- Get an adult to watch you when you do an experiment. They enjoy seeing experiments too.
- Follow the specific directions given for each experiment.
- Clean up after each experiment.

Note to Teachers, Parents, and Other Adults

Science is not merely a collection of facts, but a way of thinking. As a teacher, parent, or adult friend, you can play a key role in maintaining and encouraging a young person's interest in science and the surrounding world. As you do environmental experiments with a young person, you may find your own curiosity being expanded. Experiments are one way to learn more about the air, water, land, energy, and life upon which we all depend.

I. Energy, Temperature, and Heat

Without the sun, there would be no life on Earth. The sun warms the earth, generates wind, and carries water into the air to produce rain and snow. The energy of the sun provides sunlight for all the plant life on our planet, and through plants provides energy for all animals.

The sun is like a giant furnace in which hydrogen nuclei (atoms without electrons) are constantly smashed together to form helium nuclei. This process is called nuclear fusion. In this process, 3.6 billion kilograms (8 billion pounds) of matter are converted to pure energy every second. The temperature in the sun exceeds 15 million degrees.

Nuclear fusion is one kind of energy. Other forms of energy include: mechanical energy, heat, electrical energy, chemical energy, and light. Mechanical energy is the energy of organized motion, such as a turning wheel. Heat is the energy of random motion, such as a cup of hot water. Electrical energy is the energy of moving charged particles or electrons, such as a current in a wire. Chemical energy is the energy stored in bonds that hold atoms together. Light is any form of electromagnetic waves, such as X rays, microwaves, radio waves, ultraviolet light, or visible light.

Energy can be converted from one form to another. For example, the nuclear energy of the sun is converted to light, which goes through space to the earth. Solar collectors or mirrors can be used to focus some of that light to heat water to steam. This steam can be used to turn a turbine, which can power a generator to produce electricity.

Most of our energy needs are met by burning fossil fuels such as coal, oil, gasoline, and natural gas. The chemical energy stored in these substances is released by burning these fuels. When fossil fuels burn, they combine with oxygen in the air and produce heat and light.

Fossil fuels are not renewable. When they are used up, they are gone forever. However, renewable energy sources such as wind, sun, geothermal, biomass, and water power are renewable. They can be used over and over to generate the energy to run our society.

Tremendous amounts of renewable energy are available. For example, the solar energy that falls on just the road surfaces in the United States is equal to the entire energy needs of the country. Although there are sufficient amounts of renewable energy, we must improve our methods of collecting, concentrating, and converting renewable energy into useful forms.

In the following experiments, you will learn something about the amount of energy the sun produces at the earth's surface and how heat energy can be stored.

Experiment #1

How Much Energy
Does the Sun Produce?

Materials

Water

Paint brush

Watch or clock

Newspaper

Disposable, aluminum pie pan

Measuring cups

Black paint or spray paint (flat, not shiny)

Procedure

Go outside and spread out a sheet of newspaper. Place an aluminum pie pan on the newspaper. Carefully bend one spot on the edge of the pie pan to make a spout shape. This will allow you to more easily pour water out of the pan. Have an adult help you paint the inside of the aluminum pie pan. You can use a brush and a can of paint or spray paint. Be sure not to get the paint on anything except the disposable pie pan and the newspaper. After painting, set the pie pan where the paint can dry overnight.

You will need to do the rest of this experiment on a warm, sunny day. You do not want the pie pan to be in the shade. Set the aluminum pie pan in a warm, sunny spot. The sun will need to constantly shine on the pie pan. The black color of the pie pan allows it to absorb, rather than reflect, solar energy. You will need to begin the experiment about 11:00 A.M., so the solar heating will be done when the sun is high in the sky.

Add exactly one cup of water to the pie pan. Wait four hours while the sun is shining on the pan of water. After exactly four hours of sunshine, carefully pour the remaining water from the pie pan into a one-half or one-fourth measuring cup. Use these measuring cups to estimate the amount of remaining water to the nearest one-eighth of a cup.

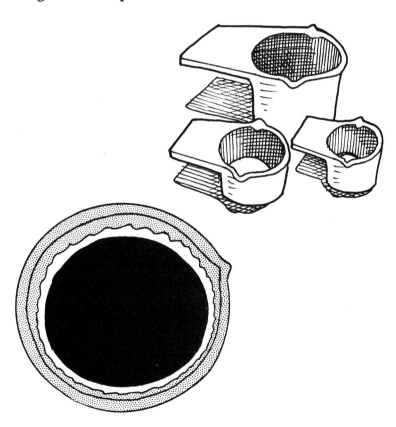

Observations

Is the amount of water in the pie pan less than when you began the experiment? How much water is left in the pie pan after four hours of sunlight?

Discussion

You will probably observe that the amount of water in the pie pan is less after four hours of sunlight exposure. Where did the water go? As the sunlight shines on the dark surface of the painted pie pan, solar energy is absorbed and heats the pan and the water. This energy causes a portion of the water to evaporate. As water evaporates, it leaves the liquid form and goes into the air as a gas (water vapor). You will probably find that some, but not all, of the water evaporated.

Scientists use the unit of joule as a measure of energy. However, you may find it helpful to think in units of dietary calories instead of joules. One dietary calorie is equal to 4,184 joules of energy. One cup of breakfast cereal with one-half cup of milk would have about 240 dietary calories, or approximately 1,000,000 joules of energy.

Although the earth receives only a tiny portion of the total energy output of the sun, the earth has a constant supply of 173 million billion (173,000,000,000,000,000) watts of solar power. A watt is a unit of power equal to a joule of energy used per second. For comparison, a typical light bulb to run a lamp in your home might require 100 watts of power. A million watts could supply the energy needs of about 500 average American homes.

Use the table below to determine the solar energy required to evaporate a certain amount of water. The amount of water remaining in the pan will allow you to determine how much energy was used, how much power was used, and the amount of power per area.

Your results will probably be in the middle range of this table. For example, if one-half of your water evaporated, then the water remaining would be one-half cup. Thus, the energy used to evaporate this water would be 289,000 joules of energy. This energy would give a power of 20 watts, and a power per area of 800 watts per square meter (watts/meter^2).

Solar Energy Required to Evaporate Water				
Water Remaining (cup)	Water Evaporated (cup)	Energy Used (joules)	Power Used (watts)	Power per Area (watts/meter^2)
1	0	0	0	0
7/8	1/8	72,250	5	200
3/4	1/4	144,500	10	400
5/8	3/8	216,750	15	600
1/2	1/2	289,000	20	800
3/8	5/8	361,250	25	1,000
1/4	3/4	433,500	30	1,200
1/8	7/8	505,750	35	1,400
0	1	578,000	40	1,600

The following procedure was used to generate the numbers in the Table. It is known that it takes 578,000 joules of energy to evaporate one cup of water. This known energy per cup is multiplied by the fraction of a cup that was evaporated. This gives the solar energy used to evaporate the water in the pie pan. The energy is divided by the number of seconds in four hours (14,400 seconds). This gives the power of the solar energy striking the pie pan, since a watt is equal to a joule per second. Finally, the power (in watts) is divided by the surface area of the pie pan (0.025 square meters) to give the power per area.

When the sun is overhead, the intensity of solar energy can be as much as 1,000 watts per square meter. If all of this energy could be converted to electricity, one square meter of sunshine would be enough to run ten 100-watt light bulbs. However, our current solar cells that convert sunlight to electricity are able to change only about 15 percent of the light to electricity.

You can see from this experiment that there is tremendous energy available from our sun. Most of this energy warms our planet or is reflected back into space. Among other things, the remaining portion of energy powers our water cycle, producing rain and snow, or provides plants with the energy they need to live.

Scientists and engineers are learning more about trapping solar energy and converting it to useful power. It has been estimated that all forms of potentially available renewable energy (wind, water, biomass, and direct solar) have an energy equivalent to 80 trillion barrels of oil. In other words, one year of renewable solar energy is 5,000 times greater than the current yearly energy needs of the United States. In comparison, it has been estimated that all the remaining coal, oil, natural gas, and other potential nonrenewable energy reserves of the United States are equal to about 8 trillion barrels of oil.

Since we do not yet know how to use a significant portion of this renewable energy, much work remains to be done. In the remainder of this book you will learn more about our current ability to use renewable energy and the promise it may hold for the future.

Other Things to Try

Repeat this experiment using different amounts of water on different days and compare the solar power you find.

Repeat this experiment in the late afternoon, when the sun is lower in the sky. How do your results compare to this experiment?

Experiment #2

How Can Water Be Used to Store Heat Energy?

Materials

Two paper cups
Measuring cups
Hot water
Watch or clock
Sink
Refrigerator (with freezer compartment)

Procedure

Turn on the hot water faucet of a sink and wait several minutes until the water is hot. Be careful not to burn yourself with this hot water. Add one-fourth cup of this hot water to the first paper cup. Add one cup of hot water to the second paper cup. Place both of these cups in the freezer compartment of a refrigerator.

After thirty minutes check the water in each cup. Return the cups to the freezer compartment and check them again after fifteen minutes. Keep checking the cups each fifteen minutes until the water in one of the cups is frozen.

Observations

Does the water in the cups freeze at the same time? Does the water in one of the cups freeze first? How long does it take for the water to freeze?

Discussion

You will probably observe that the smaller amount of water in the first cup freezes prior to the larger amount of water in the second cup. Both cups were filled with the same hot water. However, even though the water in both cups was at the same temperature, they did not freeze at the same time. The amount of heat energy stored by the water depends on both the temperature and the amount of water.

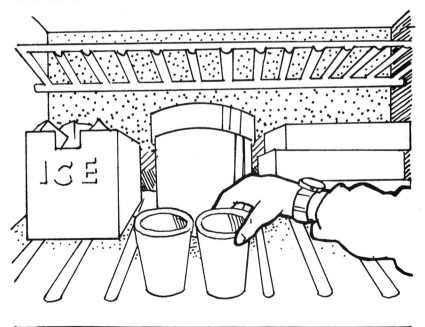

We expect that the more heat energy stored in the cup, the longer it takes the water in the cup to freeze. Since one cup of water has more heat energy than one-fourth cup of water, it takes the larger amount of water longer to freeze.

Temperature is a measure of the average hotness of an object. The hotter an object, the higher its temperature. As the temperature is raised, the atoms and molecules in an object move faster. The molecules in hot water move faster than the molecules in cold water. Remember that the heat energy stored in an object depends on both the temperature and the amount of the substance. A smaller amount of water will have less heat energy than a larger amount of water at the same temperature.

Increasing the temperature of a large body of water is one way to store heat energy for later use. A large container filled with salt water, called brine, may be used to absorb heat energy during the day when it is warm. This energy will be held in the salt water until the night when it is cooler. This stored heat energy can be released at night to warm a house or building. This is one way to store the sun's heat energy until it is needed.

Solar ponds are used to store energy from the sun. Temperatures close to 100°C (212°F) have been achieved in solar ponds. Solar ponds contain a layer of fresh water above a layer of salt water. Because the salt water is heavier, it remains at the bottom of the pond—even as it gets quite hot. A black plastic bottom helps absorb solar energy from sunlight. The water on top serves to insulate and trap the heat in the pond.

In a fresh water pond, as the water on the bottom is heated from sunlight, the hot water becomes lighter and rises to the top of the pond. This convection or movement of hot water to the top tends to carry away excess heat. However, in a salt water pond, there is no convection so heat is trapped. In Israel a series of salt water, solar ponds were developed around the Dead Sea. The heat stored in these solar ponds has been used to run turbines and generate electricity.

Other Things to Try

Place one-half cup of hot tap water in one cup. Place one-half cup of cool tap water in a second cup. Put both cups in the freezer compartment of a refrigerator and check them every fifteen minutes until the water freezes solid. Which cup of water do you think has more heat energy? Which cup of water do you think will freeze first?

Place one cup of water in one paper cup and one-fourth cup of water in a second paper cup. Put both cups in the freezer of a refrigerator and leave overnight. The next day remove both cups of frozen water. Set the two cups out in the room. Observe the time it takes each piece of ice to melt. Which piece of ice do you think will melt first? Which piece will require more heat energy to cause it to melt?

II. Sources and Savings of Energy

The United States has large reserves of coal, natural gas, and crude oil which is used to make gasoline. However, the United States uses the energy of millions of barrels of crude oil every day, and it must import about half its crude oil from other countries.

Burning fossil fuels (oil, coal, gasoline, and natural gas) produces carbon dioxide gas. Carbon dioxide is one of the main greenhouse gases that may contribute to global warming. In addition, burning coal and gasoline can produce pollution molecules that contribute to smog and acid rain.

Using renewable energy—such as solar, wind, water, biomass, and geothermal—could help reduce pollution, prevent global warming, and decrease acid rain. Nuclear energy also has these advantages, but it requires storing radioactive wastes generated by nuclear power plants. Currently, renewable energy produces only a small part of the energy needs of the United States. However, as technology improves, renewable energy should become less expensive and more common.

Hydropower (water power) is the least expensive

way to produce electricity. The sun causes water to evaporate. The evaporated water falls to the earth as rain or snow and fills lakes. Hydropower uses water stored in lakes behind dams. As water flows through a dam, the falling water turns turbines that run generators to produce electricity.

Currently, geothermal energy (heat inside the earth), biomass (energy from plants), solar energy (light from concentrated sunlight), and wind are being used to generate electricity. For example, in California there are more than sixteen thousand (16,000) wind turbines that generate enough power to supply a city the size of San Francisco with electricity.

In addition to producing more energy, we can also help meet our energy needs through conservation. Conservation means using less energy and using it more efficiently.

In the following experiments, you will use wind to do work, examine how batteries can store energy, and see how insulation can save energy.

Experiment #3

Can Wind Be Used as a Source of Energy?

Materials

Pinwheel (can be purchased or made from construction paper

Small shoe box (children's size)

Lightweight string (about four feet long)

Paper clips

Tape

Electric fan

Plastic straw (longer than the width of the shoe box)

Hole punch

Procedure

In this activity, you will try to use the energy in the wind to lift a set of six paper clips. You will first need to construct your windmill.

Use a hole punch to punch holes in the opposite sides across the width of a small, cardboard shoe box. Use the narrow sides of the box so the two holes are less than six inches (15 centimeters) apart. Make sure the holes are directly opposite each other. Place a plastic straw through the two holes. You may need to use the hole punch to enlarge the holes so the straw can rotate within the holes. The ends of the straw should extend out either side of the box.

Use the blades from a purchased pinwheel, or cut and fold a square piece of construction paper into the shape of a pinwheel. Next you will need to attach the pinwheel blades to one end of the straw. Partially unfold a small paper clip and insert the larger end into the straw. Push the straightened end of the paper clip through the center of the pinwheel. Bend this end of the paper clip and tape it to the outside of the pinwheel.

Set the electric fan on a table or countertop. Hold the shoe box so that the pinwheel is free to turn. HAVE AN ADULT PLUG IN AND TURN ON THE FAN. Move the windmill box to direct the breeze from the fan toward the blades of the pinwheel. Move the box until you find the best angle of the fan to the pinwheel so that the pinwheel turns freely and rapidly.

Turn off the fan. Now tape one end of the string to the side of the straw with no pinwheel just outside the box, and wrap the string around the straw a few times. Tie the other end of the string to a paper clip. Attach five other paper clips to the paper clip tied to the string. Allow the string to hang down so that the paper clips on the end of the string rest on the floor.

Now, you will test to see if your windmill can convert wind power to do work and lift the paper clips off the ground. Turn on the fan and hold the box where you did before to make the pinwheel turn.

Observations

Does the windmill turn the straw? Does the string wrap around the straw as the straw turns? What happens to the paper clips?

Discussion

You should observe the straw shaft turning as the wind from the fan is directed toward the blades of the pinwheel. As the pinwheel turns, it should wrap the string around the straw and lift the paper clips into the

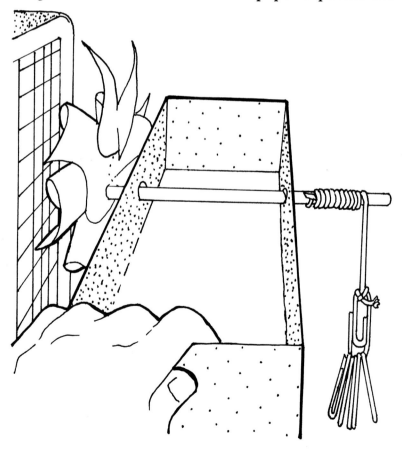

air. Your windmill converts the energy of the wind to work and lifts the weight of the paper clips. If your windmill is not working, then examine all the parts. Compare your setup to the drawing, and see if any changes need to be made in your construction.

One way to store the energy produced by a windmill is to lift a weight. When the weight is allowed to fall, work can be produced. Weights in a grandfather clock are used to store energy and can run a clock for a week or longer. A windmill's energy can be used to pump water to a storage area at a higher elevation. Later, this water can be allowed to fall through a turbine which turns a generator and produces electricity.

Electricity can also be produced directly from wind power. The shaft, or rod to which the windmill blades are attached, can be used to turn a generator. A generator or dynamo is used to convert mechanical energy into electrical energy. Power conversion units can change the direct current that wind generates to an alternating current. The alternating current can be fed directly into utility lines and used in our homes.

The sun is the original source of wind power. Without the sun to heat the earth, there would be no wind. The energy of the sun heats the earth, but all parts of the earth are not at the same temperature. These differences in temperature are responsible for global and local patterns of wind. For example, during the day a constant wind blows from the sea toward the land along coastal regions.

Air above the hotter land rises and cooler, heavier air above the ocean moves in to take its place.

The power of the wind can be harnessed to do work. For at least 4,000 years, the wind has been used to move sailing ships. The wind has enough power to move ships across oceans and around the world.

For at least a 1,000 years, windmills have been used for pumping water and turning stones to grind grain. Millions of windmills have been used on the plains of America, Africa, and Australia to pump water from deep wells for livestock and humans.

In this century, windmills or wind engines have been used to generate electricity. Over 15,000 wind engines were installed in California in the 1980s. These wind engines have the capability to produce up to 1.5 billion watts of electricity. In California in 1987, wind was used to produce as much electricity as the city of San Francisco uses in an entire year.

Other Things to Try

Repeat this experiment and find the maximum weight you can lift with your windmill. Try more paper clips or try a heavier weight such as a pen.

List some of the problems associated with using windmills. What happens when the wind is blowing too gently? What happens if the wind blows strongly, such as in a storm? Do you think the area where you live is windy enough for wind engines to produce electricity?

Experiment #4

Can a Battery Be Used to Store Energy?

Materials

Earphone or headset for a portable radio

Two wires with alligator clips on each end of the wires

AA-size battery

Spoon

Small piece of aluminum foil

Tomato juice

New, shiny penny

Plate

Procedure

Examine the metal shaft of the part of the earphone or headset that is inserted into a portable radio. You will notice that just below the tip of the shaft there is a plastic spacer. Clip on one of the wires below this spacer. Then clip on the other wire above this spacer.

To test that the wires are properly connected to the earphone or headset, take the unconnected ends of the two wires and touch them to an AA-size battery. One wire should touch the positive end of the battery, while the other is touching the negative end of the battery. Place the earphone or headset to your ear. If your connections are made correctly, you should hear a crackling sound in the earphone or headset. If you do not hear a crackling sound, check your connections carefully.

Place a small piece of aluminum foil, about five inches (13 centimeters) square, on a small plate. Using a spoon, make a puddle of tomato juice on the aluminum foil. The puddle of tomato juice should be slightly larger than a penny. Next, place a new, shiny penny face down in the puddle of tomato juice.

Using the alligator clip, attach one of the wires connected to the earphone to one of the edges of the aluminum foil. Take the end of the other wire and touch the alligator clip to the penny. Move the alligator clip over the penny.

Observations

Do you hear a crackling sound when you touch the alligator clips to the penny in the puddle of tomato juice? What do you hear when you move the alligator clip over the penny? What do you hear when you stop touching the penny with the alligator clip?

Discussion

A battery is a device that produces electrical energy from a chemical reaction. Another name for a battery is voltaic cell. Voltaic means to make electricity.

Most batteries contain two or more different

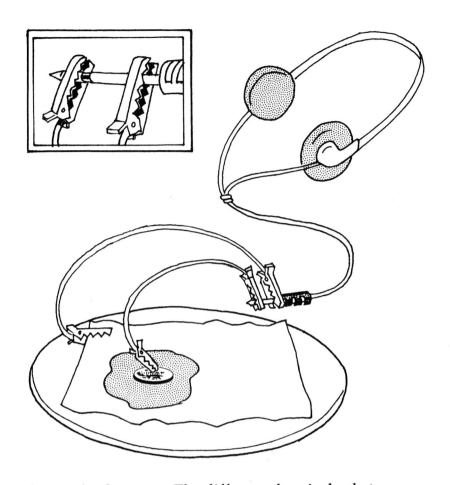

chemical substances. The different chemical substances are usually separated from each other by a barrier. One side of the barrier is the positive terminal of the battery and the other side of the barrier is the negative terminal. When the positive and negative terminals of a battery are connected to a circuit, a chemical reaction takes place between the two different chemical substances that produces a flow of electrons (electricity).

When a battery is producing electricity, one of the chemical substances in the battery loses electrons. These electrons are then gained by the other chemical substance.

A battery is designed so that the electrons lost by one chemical substance are made to flow through a circuit, such as a flashlight lamp, before being gained by the other chemical substance. A battery will produce a flow of electrons until all of the chemical substances involved in the chemical reaction are completely used.

In this experiment you made a simple battery with a penny, aluminum foil, and tomato juice. You completed a circuit with your battery by touching one of the wires attached to the earphone or headset to the penny, while touching the other wire to the aluminum foil. When you completed the circuit, a flow of electrons was produced by your battery. The crackling sound you heard was caused by the earphone or headset converting electrical energy from your battery into sound energy.

In your battery, the aluminum in the aluminum foil loses electrons. The other part of the reaction is more complex. Either the acid in the tomato juice or copper ions (that form when the copper metal in the penny reacts with the acid in the tomato juice) gain the electrons lost by the aluminum.

The main types of batteries are known as primary and secondary batteries. Dry cell batteries, like the ones used in flashlights and portable radios, are primary batteries. Another important primary battery is the mercury battery. Mercury batteries are typically small and flat. They are used to power cameras, watches, hearing aids, and calculators.

An advantage of primary batteries is that they are generally inexpensive. One disadvantage is that they cannot be recharged. When the chemical substances in the primary batteries are used up, the battery is dead.

Lead storage batteries and nickel-cadmium (NiCad) batteries are examples of secondary batteries. Car batteries are lead storage batteries. Flashlight batteries that are rechargeable are NiCad batteries. Secondary batteries are more expensive than primary batteries. However, unlike primary batteries, lead storage batteries and NiCad batteries can be recharged repeatedly.

Other Things to Try

Repeat this experiment using other coins such as a dime, nickel, or quarter. Do any of these coins cause a louder crackling sound in the earphone or headset?

Repeat this experiment using a nail instead of a coin.

Can you make a battery with other juices? To find out, repeat this experiment with other juices such as lemon and orange juice. What do you observe?

Experiment #5

Can Insulation Save Energy?

Materials

Two large, plastic cups

Ice

Watch or clock

Small, insulated ice chest or cooler (Styrofoam or plastic)

Procedure

Completely fill each plastic cup with ice. Set one plastic cup out in the room. Set the other plastic cup of ice inside the small ice chest or cooler. Close the lid of the ice chest and leave both cups undisturbed.

Observe the cup of ice left out in the room once every hour. When all the ice has changed to water in the cup left out in the room, open the smaller cooler and observe the cup of ice inside. Continue to check on the cup of ice in the cooler about once an hour and see how long it takes to melt.

Observations

How long does it take for the cup of ice in the room to melt and change to water? When all the ice has melted in the cup in the room, is there still ice in the cup in the ice chest? How long does it take for the ice in the cooler to melt?

Discussion

You should find that the ice in the room melts faster than the ice inside the ice chest. When the ice in the room has all melted and only water remains, the cup inside the insulated cooler may still be mostly ice. Even though both cups were filled with the same amount of ice, they did not melt at the same time.

The temperature in the room is much warmer than the temperature of the cold ice. Since heat always flows from a higher temperature to a lower temperature, the heat in the room flows or moves into the ice and causes it to melt.

35

An insulator is a substance that partly blocks or slows the flow of heat through it. Styrofoam is a lightweight plastic used in drinking cups. Styrofoam is a good insulator. A cooler or ice chest that is made of Styrofoam or some other insulator tends to block the flow of heat through it.

Heat flows into buildings during warm summer months and from buildings during cold winter months. Energy must be used to cool buildings in the summer and heat them in the winter. Since insulation can slow the flow of heat, the use of insulation in buildings can save energy.

Some common home and building insulation materials include Styrofoam, polyurethane foam, and fiberglass. These materials are all good insulators, which means that they are poor conductors of heat. Placing these insulating materials on attic floors or in building walls tends to trap heat inside during the cold winter and keep heat out during the hot summer.

Plastic foams filled with trapped gas tend to block heat flow. The chemicals used to make polyurethane foam can be sprayed directly into the spaces between walls. These chemicals produce carbon dioxide gas and polyurethane plastic. The gas tends to spread the polymer apart so the weight is mostly plastic but the volume is mostly trapped gas. Polyurethane also is used to insulate refrigerators, refrigerated trucks, pipes, and building walls.

Fiberglass insulation is frequently used in attic floors to insulate homes. Also, fiberglass insulation is used to insulate the Trans-Alaska pipeline. This pipe carries oil 800 miles from Prudhoe Bay in northern Alaska to Valdez in southern Alaska. The crude oil that travels through this pipe is easier to pump if it is hot. An insulated pipeline requires less energy to keep the oil hot.

Energy conservation becomes more and more important as energy costs rise. A great deal of energy is used to cool buildings in summer and heat buildings in winter. Less energy will be needed if buildings are well insulated and energy is not wasted.

Other Things to Try

Repeat this experiment with a larger amount of ice, and you may see a greater difference between insulated and uninsulated containers. Try filling a bucket and an ice chest with equal amounts of ice, and set both containers outside on a warm day. How long does it take the ice in the bucket to melt? How long does it take the ice in the ice chest to melt?

You can reverse the experiment by comparing the flow of heat from the inside of a cup to colder surroundings. Use hot water from a sink faucet to partially fill a Styrofoam cup and a glass cup. Place the Styrofoam cup inside two other Styrofoam cups and cover the top with several layers of aluminum foil. Next place the Styrofoam cups and the glass cup inside the refrigerator and compare how long it takes for the water to cool.

Fill a thermos bottle with hot water. Wait about four hours and check the temperature of the water. Continue to check the temperature about every two hours to find out how long the thermos bottle can keep the water warm. A thermos bottle has glass walls with a vacuum between them and silvered surfaces to reflect heat. The vacuum in a thermos makes an excellent insulator because there are few gas molecules to transfer heat.

III. Cooling and Heating

Cooling and heating are opposite processes. Cooling is the removal of heat energy from an object or space and heating is the addition of heat energy to an object or space. We use these opposite processes a great deal in our daily lives. For example, in the kitchen we use the cooling provided by a refrigerator to keep food cold. We also use the heat from a stove to cook food.

Nearly 75 percent of the energy used by the average family household in the United States goes for cooling and heating purposes. Air conditioning and refrigeration are the major cooling requirements of a home, while water and space heating are the most important heating requirements.

In the experiments that follow you will learn more about cooling and heating. You will also learn alternative ways of cooling and heating, using such unusual materials as gases, salts, water, and trees.

Experiment #6

Can a Salt Be Used for Heating and Cooling?

Materials

Epsom salt

Water

Two small, zipper-close, plastic bags

Measuring cups

Aluminum pie pan

Oven

Insulated mitt

Sink

Procedure

ASK AN ADULT TO HELP YOU WITH THIS EXPERI-MENT. DO NOT USE THE STOVE BY YOURSELF.

Ask an adult to turn on the oven and to set the temperature of the oven to 450° F (232° C). Pour one-half cup of Epsom salt into an aluminum pie pan and gently shake it to evenly spread the Epsom salt over the bottom of the pan. Place the pie pan in the oven. Heat the Epsom salt in the hot oven for thirty minutes.

Ask an adult to remove the pan of Epsom salt from the hot oven using an insulated mitt. Place the pan on the stovetop and allow it to cool for ten minutes. Make sure the oven is turned off.

Add one-quarter cup of the Epsom salt that was not

heated in the oven to a small plastic zipper-close bag. Next add one-quarter cup of room temperature water to the bag, seal, and shake the bag. Feel the temperature of the outside of the bag.

Next add one-quarter cup of the cooled Epsom salt that was heated in the oven to the second zipper-close bag. Add one-quarter cup of room temperature water to the bag and seal the bag. Give the bag a couple of shakes and then feel the outside of the bag.

When you are finished with this experiment, pour the contents of both bags down a sink drain. Then flush the bags and the sink with water. Also rinse out the aluminum pie pan with water. DO NOT DRINK ANY OF THE LIQUID AND DO NOT EAT ANY OF THE EPSOM SALT. EPSOM SALT CAN MAKE YOU SICK IF YOU EAT IT.

Observations

Do you notice any difference in the appearance of the Epsom salt after it is heated in the oven? Does the bag containing the water and the Epsom salt that was not heated feel warm or cool after shaking? Does the bag containing the water and the Epsom salt that was heated feel warm or cool after shaking?

Discussion

Epsom salt is a hydrate of the salt called magnesium sulfate. A hydrate is a chemical substance containing

water combined with another chemical substance (usually a salt). The water molecules in a hydrate are called waters of hydration. They can usually be removed from the hydrate by heating. The process of removing water from a hydrate is called dehydration.

In this experiment, when you heated the Epsom salt in the oven, you removed most of the water molecules (waters of hydration) in the salt.

A salt is made of positive and negative ions. Ions are charged atoms or groups of atoms. In Epsom salt, magnesium ions are positive and sulfate ions are negative. The waters of hydration surround these ions in Epsom salt.

You should find that the bag containing water and the dehydrated Epsom salt feels warm. The bag containing water and the hydrated Epsom salt feels cool. When dehydrated Epsom salt is mixed with water, heat is given off. When hydrated Epsom salt is mixed with water, heat is absorbed. When something gives off heat it feels warm, while something that absorbs heat feels cool.

Energy is required to remove individual ions from a salt crystal. However, energy is given off when the individual ions that break away from the crystal become surrounded by water molecules that are dissolving the salt. If more energy is required to remove individual ions from a salt crystal than is given off when the ions become surrounded by water, then the salt solution becomes cold. If more energy is given off when the ions become surrounded by water than is needed to remove individual ions from a salt crystal, then the salt solution becomes warm.

When hydrated Epsom salt dissolves in water, more energy is required to remove the magnesium and sulfate ions (and the waters of hydration) from the crystals than is given off when the magnesium and sulfate ions become surrounded by the dissolving water molecules.

This is why the bag containing the unheated Epsom salt became cold when you added water.

On the other hand, when dehydrated Epsom salt dissolves in water, more energy is given off by the ions becoming surrounded by water molecules than is needed to break the magnesium and sulfate ions from the crystals. This is why the bag containing the dehydrated Epsom salt became warm when you added water.

Most instant hot packs and cold packs that are available in drugstores work on the principle just discussed. When the cold or hot pack is needed, the bag is squeezed to cause the water and salt to mix. Depending on the salt used in the pack, energy is either absorbed (cold pack) or given off (hot pack). Ammonium nitrate is the most commonly used salt in cold packs. And calcium chloride is the most commonly used salt in hot packs.

Other Things to Try

Repeat this experiment using a thermometer to measure the temperature change when the Epsom salt is dissolved in water. Use the thermometer to measure the temperature change when dehydrated magnesium sulfate is dissolved in water.

Repeat this experiment using table salt. Do you observe a temperature change when the table salt dissolves in water?

Experiment #7

How Can Shade Trees
Keep Things Cool?

Materials

Two large, glass jars   Shady tree
Water   Thermometer
                                 (optional)

Procedure

You will want to do this experiment on a warm, sunny day. Fill both large, glass jars nearly full with water. Place one of the jars outdoors in a spot where sunlight will strike it for several hours. Place the other jar outdoors under a shady tree.

After three hours, feel the water in the jar that was left in the sun. Next feel the water in the jar kept under the shady tree. If you have a thermometer, check the temperature of the water in each jar. Also check the air temperature with the thermometer.

Observations

Does the water in the jar left in the sun feel warmer or cooler than the water in the jar left under the shady tree? If you are using a thermometer, what is the temperature

of the air and of the water in each jar? What is the
temperature difference of the water in the two jars?

Discussion

The water in the jar that was left in the sun should feel
much warmer than the water in the jar left under the
shady tree. The temperature difference of the water in
the two jars will vary depending on the air temperature

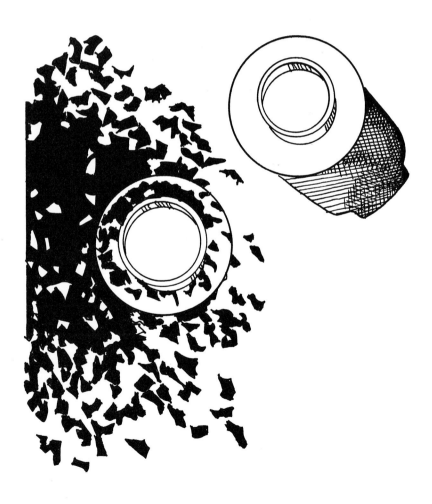

and wind conditions. On a hot, sunny day you may find that the temperature of the water in the jar left in the sun is over 18° F (10° C) warmer than the air temperature and the temperature of the water in the jar kept in the shade.

The jar of water left in the sun is constantly bathed with light energy from the sun. Some of this light energy is absorbed by the glass jar and the water in the jar. The light energy that is absorbed by the glass jar and the water is converted into heat energy. This is why the jar of water becomes warm after being in the sun for several hours.

The tree shades the other jar of water from the sun. The jar of water under the tree does not absorb light energy directly from the sun as the other jar did. Instead, much of the light energy from the sun is absorbed and used by the tree.

Since the jar of water under the tree does not absorb light energy directly from the sun, it remains cooler than the jar of water kept in the sun. You may even find that the temperature of the water in the jar shaded from the sun is a few degrees cooler than the air temperature. This is caused by water evaporating from the jar. Heat energy is removed when a liquid evaporates, and the liquid becomes slightly cooler.

Having shade trees around a house can decrease the cost of cooling the house with air conditioning. A house not shaded from the sun absorbs some of the light from the sun and heats up the outside surface of the house. If the house is poorly insulated, some of this heat will penetrate into the house, heating up the inside. The air conditioner will use more energy to remove this added heat.

Properly designed roof overhangs can significantly decrease the heating and cooling costs of a house. Because the earth's axis is tilted, the sun is lower in the winter in the northern hemisphere. In the summer, the sun is higher in the sky. A properly designed roof overhang allows sunlight in the winter to shine through windows and warm the furnishings in the rooms that receive the direct sunlight. This reduces the heating cost in the winter. In the summer, the overhang blocks the sunlight from shining into the window and heating the furnishings. This reduces the cooling cost in the summer.

Other Things to Try

Repeat this experiment at different times of the day. Do you get similar results?

Repeat this experiment when it is cloudy. How does the difference in the temperatures of the jars of water compare on sunny and cloudy days?

Experiment #8

Can the Evaporation of Water Be Used for Cooling?

Materials

Unglazed clay flower pot

Unglazed clay saucer

Modeling clay

Plastic bowl

Water

Thermometer (optional)

Procedure

Make sure to use an unglazed clay flowerpot for this experiment. Most unglazed clay flowerpots have an orange-red color and rough surfaces. The bottom of the unglazed clay saucer should have a diameter larger than the top diameter of the flowerpot. (Flowerpots and saucers can be purchased at most hardware and plant stores.)

Most clay flowerpots have drainage holes. Check to see if your clay flowerpot has a drainage hole in the bottom. If the one you are using has a hole, seal it with a piece of modeling clay. Add some water to the pot to make sure your seal is watertight.

Place the clay flowerpot and plastic bowl outside on a hot, sunny day and in a spot where they will get plenty of sunshine. Fill both the clay flowerpot and the plastic bowl nearly full with water. Place the clay saucer on top of the flowerpot and fill it with water.

Every hour for three hours, place your hands around the outside of the plastic bowl of water. Then place your hands around the outside of the clay pot filled with water. Next dip your hand into the plastic bowl of water. Remove the clay saucer from the flowerpot and dip your same hand into the pot of water. If you have a thermometer, place it in the plastic bowl of water. Read the temperature on the thermometer after it has been in the water thirty seconds. Next place the thermometer in the clay pot of water. After thirty seconds read the temperature of the water in the clay pot.

Observations

Does the outside of the clay flowerpot appear wet and feel wet after it has been outside several hours?

Does the outside of the plastic bowl feel warmer or cooler than the outside of the clay flowerpot after both containers have been in the sun several hours? Does the water in the clay flowerpot feel warmer or cooler than the water in the plastic bowl?

If you are using a thermometer, what is the temperature of the water in the plastic bowl? What is the temperature of the water in the clay pot?

Discussion

You should find that soon after you add the water to the unglazed clay flowerpot, the outside of the clay pot looks and feels wet. The outside of the clay flowerpot should feel cooler than the outside of the plastic bowl. Also, the water inside the flowerpot should feel much cooler than the water inside the plastic bowl. In fact, if you measured the temperatures of the water in both containers you may have found that the water in the clay pot was cooler by as much as 18°F (10°C). The temperature difference will depend on the air temperature, the humidity, and wind. The higher the humidity and the less wind there is, the smaller the temperature difference between the two containers will be.

The outside of the clay flowerpot becomes wet because water inside the flowerpot moves through the wall of the clay pot. When water reaches the outside surface of the clay pot, it can evaporate. When the water evaporates, it changes from a liquid to a gas or vapor. Heat energy is required for water to evaporate. This heat energy comes from the clay pot and the water inside the clay pot. When the water evaporates and removes heat from both the pot and water, the temperature of both the pot and water decreases.

Water evaporates from the plastic bowl as well. However, since water cannot move through the plastic, water evaporates only from its surface in the plastic bowl. More heat is removed from the clay pot than the plastic bowl because water is evaporating from a larger area. This is why the outside of the clay pot and the water inside it should feel cooler than the outside of the plastic bowl and the water inside it.

The evaporation of water for cooling purposes is called evaporative cooling. An important example of this type of cooling is the removal of body heat by humans through sweating. When your body needs to cool, perspiration is released to the surface of your skin where it evaporates. The evaporation of the water in the perspiration causes your skin to cool.

Breezes feel particularly cooling when you have perspiration on your skin. This is because the increased movement of air over your body evaporates more water from your skin than still air does. Water on your skin evaporates more slowly when the humidity is high. This is because the humid air already contains much water vapor. Humid air absorbs less water as vapor than dry air.

Electrical power plants that burn fossil fuels or use nuclear energy to generate electricity use huge water cooling towers for cooling purposes. The water to be cooled is pumped to the top of the tower and allowed to drip down through the tower. As the water moves down the tower, air from the bottom of the tower moves up through the tower, evaporating some of the falling water. The heat lost by the evaporating water cools the remaining water that is collected in a basin under the tower. One pound of water that evaporates in a tower can lower the temperature of 100 pounds (45 kilograms) of other water by nearly 5° C (10° F).

Other Things to Try

Repeat this experiment and record the temperatures of the water in the two containers with a thermometer every hour for five or six hours. Is the temperature of the water in the clay pot always lower than the temperature of the water in the plastic bowl?

If you are ever near the bottom of a waterfall, notice that the air temperature around the waterfall is cooler than air away from the waterfall. Can you explain why?

Experiment #9

Can Expanding Gases Be Used for Cooling?

Materials

Thermometer (outdoor type)

Newspaper

Can of Lysol® spray (use pressurized aerosol can, not pump spray)

Procedure

HAVE AN ADULT HELP YOU WITH THIS EXPERIMENT. LYSOL®, LIKE OTHER SPRAY CANS, SHOULD NOT GET NEAR A FLAME, AND THE CONTENTS SHOULD NOT GET IN YOUR EYES.

Set a can of Lysol® spray in the room where the experiment will be done and wait one hour. Observe the temperature of the room by reading the temperature on a thermometer. The can of Lysol® spray should be the same temperature as the room.

Place about ten sheets of newspaper, one on top of the other, on the floor. Have an adult hold the thermometer in one hand and the spray can in the other hand. These items should be held above the newspaper to catch Lysol® liquid that may drip off the thermometer. The adult should hold the nozzle of the

Lysol® spray can about four inches from the bottom of the thermometer. Have the adult spray the bottom of the thermometer for twenty seconds.

While the adult sprays the Lysol® on the thermometer, you should observe the temperature. After he or she stops spraying, continue to observe the temperature on the thermometer for several minutes.

Observations

What is the temperature of the room and the can of Lysol® spray? Does the temperature drop while the Lysol® is sprayed? What is the temperature while gas is sprayed from the can on the thermometer? What happens to the temperature after the spraying is stopped?

Discussion

You should find that the temperature on the thermometer drops as the Lysol® is sprayed on it. The temperature may decrease by 14° F (8° C) or more. After the spraying is stopped, the temperature may drift slightly lower, and then should gradually increase to the original temperature of the room.

If you shake the can of Lysol®, you should hear the sound of a liquid sloshing or moving inside the can. The can contains gas and liquid. When the can is sprayed, the pressurized gas escapes and liquid expands to a fine mist or vapor (gas). This change, from a higher pressure

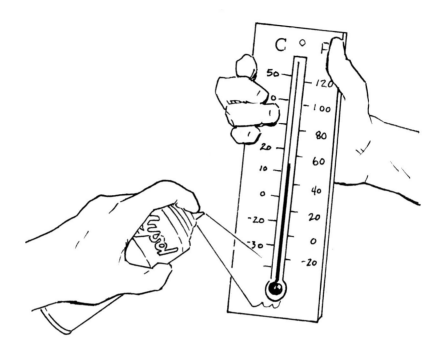

to a lower pressure gas and from a liquid to a gas, causes
a cooling effect. You observe this cooling when you see
the thermometer's temperature decrease.

In a typical air conditioning or refrigeration system,
a liquid at high pressure is allowed to pass through a
valve from a higher pressure to a lower pressure. As the
liquid enters the lower pressure region, it changes from
a liquid to a gas. This change causes a cooling effect. The
liquid cools as it changes to a gas.

In a cooling system, such as a refrigerator or air
conditioner, this cold gas is used to cool a box
(refrigerator) or a room (air conditioner). Then the cool
gas is forced through a compressor pump where it
undergoes a warming effect and changes back to a

liquid. This excess heat is removed before the liquid is expanded to a gas again. In an air conditioner, the excess heat is blown outside.

Special molecules containing chlorine, fluorine, and carbon atoms are used in most cooling systems. These Freon or chlorofluorocarbon (CFC) molecules are used because they are stable, nontoxic, and will not burn.

In recent years, scientists have discovered that these Freon or CFC molecules are damaging the earth's ozone layer. Ozone molecules in the upper atmosphere block harmful ultraviolet radiation from reaching the earth. Because these CFC molecules are so stable they tend to stay in the atmosphere for many years, during which time they gradually spread to the upper atmosphere.

In the upper atmosphere, CFC molecules can release chlorine atoms. These atoms cause a chemical reaction that breaks apart ozone. One chlorofluorocarbon molecule may destroy thousands of ozone molecules. Scientists and engineers are looking for new methods of cooling and new gases that are less damaging to the ozone layer.

The main energy used in operating a cooling system is the energy required to run a compressor to force a gas to a higher pressure, where it will change back to a liquid. This energy is normally supplied by electricity or by burning natural gas to run a compressor pump. However, there are systems in which solar energy is used to supply the energy needed for cooling.

Other Things to Try

Open a two liter soft drink bottle that has been out in the room for several hours. As you open the bottle, let the carbon dioxide gas that escapes and causes a hissing sound fall on your lips. Does the gas coming out of the soft drink bottle feel hot or cold?

IV. Solar Energy

The energy of sunlight powers our biosphere (air, water, land, and life on the earth's surface). About 50 percent of the solar energy striking the earth is converted to heat that warms our planet and drives the winds. About 30 percent of the solar energy is reflected directly back into space. The water cycle (evaporation of water followed by rain or snow) is powered by about 20 percent of the solar energy.

Some of the sunlight that reaches the earth is used by plants in photosynthesis. Plants containing chlorophyll use photosynthesis to change sunlight to energy. Since these green plants form the base of the food chain, all plants and animals depend on solar energy for their survival.

When the sun is overhead, about 1,000 watts of solar power strike 1 square meter (10.8 square feet) of the earth's surface. Using solar cells, this solar energy can be converted to electricity. However, because sunlight cannot be converted completely to electricity, it takes at least a square meter of area to gather enough sunlight to run a 100-watt light bulb.

Solar energy is still more expensive than other methods of generating electricity. However, the cost of solar electricity has greatly decreased since the first solar cells were developed in 1954.

It has been proposed that panels of solar cells on satellites in orbit above the earth could convert solar energy to electricity twenty-four hours a day. These huge solar power satellites could convert electrical energy to microwaves and then beam these microwaves to Earth. At the earth's surface, tremendous fields covered with antennas could convert the microwave energy back to electricity.

It would take thousands of astronauts many years to build such a complicated system. However, there are many practical uses of solar energy in use today. These uses include heating water, heating and cooling buildings, producing electricity from solar cells, and using rain and snow from the water cycle to power electrical generators at dams.

In the following experiments, you will examine the use of solar energy in heating water, cooking foods, concentrating sunlight, and producing electricity.

Experiment #10

Can the Sun Be Used to Heat Water?

Materials

Paint brush

Newspaper

Water

Large, plastic glass

Empty aluminum 12-ounce (355 milliliter) soft drink can

Black paint or spray paint (flat, not shiny)

Thermometer (outdoor type)

Aluminum foil

Procedure

Go outside and spread a sheet of newspaper on the ground. Place an empty aluminum soft drink can on the newspaper. Have an adult help you paint the outside of the aluminum can. You can use a brush and can of paint or spray paint. Be sure to use paint that is suitable for a metal surface. The paint should give you a flat (not shiny) surface. Be sure not to get the paint on anything but the can and newspaper. After painting, set the can where the paint can dry overnight.

You will need to do the rest of this experiment on a warm, sunny day. Partially fill a large, plastic glass with cool tap water. Check the temperature of the water with a thermometer. Pour the water from the plastic glass into the painted black can, completely filling the can. Pour out any extra water remaining in the plastic glass. Cover the can's opening with a small piece of aluminum foil about the size of a quarter.

Set the black can outside in a sunny spot. Pick a place where the sun will shine on the can all day. (You do not want the can to be in the shade.)

After the can of water has been in the sunshine for about four hours, pour the water into the large, plastic glass. Check the temperature of the water with the thermometer. Feel the outside of the can.

Observations

What was the temperature of the cool tap water when it was first placed into the black can? What is the temperature of the water after it was heated in the can for four hours? Does the outside of the can feel hot?

Discussion

You should find a significant increase in the temperature of the water that was left in the black can during the day. The tap water initially may be about 21°C (70°F), but after the water has been heated inside the can, the temperature should rise to more than 38°C (100°F). The exact temperature you achieve in your miniature, solar water heater (black can) will depend on your location and the time of year. However, you

should find that the water temperature will go much higher than the temperature of the outside air.

The electromagnetic radiation from the sun includes ultraviolet, visible, and infrared radiation. Ultraviolet radiation is the type of sunlight that causes tanning of skin. Visible radiation is the type of sunlight we see with our eyes. Infrared radiation is the type of sunlight that we feel as heat when the sun is shining on our skin. All these forms of solar radiation have energy associated with them.

When solar energy from the sun's electromagnetic

radiation strikes a black surface, solar energy is converted to heat energy and the surface is warmed. Other colors will absorb solar energy, but lightly colored surfaces tend to reflect the light, while darker colors absorb the solar energy. You may have noticed this difference if you ever walked barefoot on a dark road on a hot summer day.

Direct solar energy is not hot enough for cooking. The higher temperatures required for cooking or for changing water to steam require concentrating the energy of sunlight with mirrors or lenses. However, directly absorbed solar energy is hot enough for heating homes and producing hot water with little or no energy costs.

When we turn on a hot water faucet at a sink, water is taken from a hot water tank. In industrialized countries, we usually heat water using electricity or natural gas and store the hot water in this insulated tank. However, around the world, there are millions of solar heaters used for heating water.

Solar water heaters use a black metal plate covered with insulated glass. These solar heaters are usually placed on rooftops to receive the maximum amount of sunlight. Water flows through tubes beneath the black metal plates. Solar energy heats the black metal plates and the water passing in tubes underneath the plates. The heated water is piped to a storage tank, where it is kept until needed. If the location of the solar heater is not consistently sunny, then an auxiliary heater—using electricity or natural gas—is sometimes used to heat the water.

Other Things to Try

Repeat this experiment covering the black can with a large glass jar. This glass should help trap the heat and make the water get even hotter. What is the maximum temperature you can get with this type of solar heater?

Repeat this experiment and check the temperature of the water each half hour for three hours during the middle of the day. Quickly check the temperature of the water, then return the water to the black metal can. Write down the time that you checked the water, and next to the time, write down the temperature. How long does it take for the water to reach the maximum water temperature?

Repeat this experiment with an unpainted can and a painted can. After one hour, how do the temperatures of the water in each can compare?

Repeat this experiment on a cloudy day. Does the water in the can get as hot without sunshine?

Experiment #11

How Can the Sun Be Used for Cooking?

Materials

Three clear, clean plastic cups

Aluminum foil

Measuring cup

Two spoons

Two small tea bags

Watch or clock

Water

White sheet of paper

Plastic pan (4 inches deep and 12 inches across is a convenient size but other sizes can be used)

Procedure

You will need to do this experiment on a warm, sunny day.

Use two sheets of aluminum foil and place them crosswise to completely cover the bottom and sides of a plastic pan. Try to arrange the aluminum foil so that it is smooth and curved like a bowl. The aluminum foil will help to reflect the solar energy and concentrate the light and heat toward the center of the pan. Place this aluminum covered pan outside in a warm, sunny spot where the sunlight will shine directly on it.

Add one cup of water to each of two plastic cups. The water you add to the cups should be neither hot nor cold, but about room temperature. Place one cup of water in the middle of the pan. Turn the empty plastic cup upside down and place it on top of this cup. Leave this "solar cooker" undisturbed for one hour. The other cup of water should remain inside.

After one hour, gently place one tea bag in each of the water-filled cups. Wait ten minutes and then lift the tea bag out of each cup. Using a spoon, stir each cup of tea. Place both cups of tea on a white piece of paper and look down on the two cups to compare their darkness. Put your finger in each cup of tea to compare their temperatures.

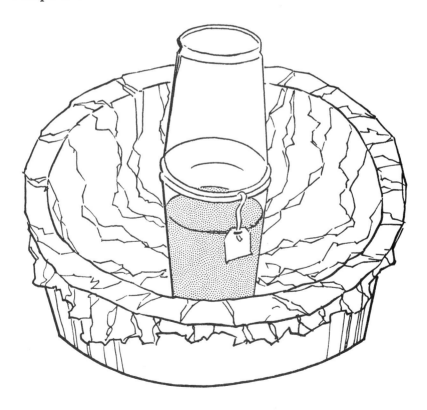

Observations

Which cup of tea is a darker color? Which cup of tea is warmer?

Discussion

You should find that the water left in the "solar cooker" is darker and warmer than the water left in the shade. The darker color indicates that more tea has gone into or dissolved in the warmer water.

Cooking involves heating food to bring about chemical changes. Sometimes foods are heated simply because the food tastes better warm than cold. In making tea, we sometimes heat water to help dissolve tea or help dissolve sugar if the tea is sweetened.

Normally the water used to make tea is heated on a range top or in a microwave oven. Using a range or microwave oven requires buying energy in the form of electricity or natural gas. Using a solar cooker does not require any energy costs because it uses a freely available renewable energy source—the sun.

A curved mirror in a bowl-like shape can focus reflected sunlight at a spot for cooking. A mirror about 1.5 meters (5 feet) across can generate a temperature of 177° C (350° F) and boil a liter of water in about fifteen minutes. In sunny areas of the world, solar cookers can be used instead of burning firewood for cooking.

Another way reflected and focused sunlight is used is to generate electricity. In southern California in 1982, a solar-thermal plant was built that can generate ten million watts of electrical power. This plant consists of 1,818 mirrored heliostats. A heliostat is a device that moves to track the sun across the sky and to reflect the sunlight at the same point. Each heliostat has twelve mirrors, and all the heliostats reflect sunlight to the same spot. The reflected light is directed at the top of a 90-meter (295-foot) tall tower. The concentrated sunlight is used to boil water and heat the steam up to 560° C (1,040° F). The steam turns a turbine that powers a generator to produce electricity.

One obvious disadvantage of solar-thermal plants is that they only operate when the sun is shining. The heat energy can be stored for a time by heating up a liquid or melting salt. Or the energy can be used to break water into hydrogen and oxygen. The hydrogen can then be stored and burned later to produce water and release energy.

Other Things to Try

Place your "solar cooker" in the sun as in this experiment, but place one plastic cup upside down in the middle of the pan. Put a pat of margarine or butter on top of this cup. Will the sun melt this butter? How long does it take to melt? Repeat this activity with a piece of soft cheese and determine if the solar heater will melt the cheese. In a more carefully made solar cooker, the reflective surfaces are angled to focus a large amount of sunlight at one spot and the temperatures obtained are much higher than in your cooker.

Set one cup of water in your "solar cooker." Set a second cup of water in the sunshine and leave both cups for one hour. Use a thermometer to check the temperature of each cup of water. Does your "solar cooker" help focus the sun's rays and increase the temperature?

Experiment #12

Can Solar Energy Be Concentrated?

Materials

Lamp with a single incandescent bulb

Magnifying lens

Procedure

The results of this experiment may be easiest to observe if done at night in a dark room.

Ask an adult to remove the lamp shade from a lamp that uses a single incandescent bulb. An incandescent bulb is the type that gets quite hot when used. Turn on the lamp. Turn off all the other lights in the room.

Stand about two feet from the wall that is the greatest distance from the lamp. There should be nothing between you and the lamp bulb. Place the magnifying glass on the wall so that the lens is flat against the wall. Now, slowly move the lens away from the wall and toward the light. Keep the lens parallel to the surface of the wall. As you move the lens outward, watch the wall.

Observations

Does an image of the lamp appear on the wall? How bright is this image? How big is this image?

Discussion

You should see an upside down image of the light bulb appear as you move the magnifying lens away from the wall. The image should be much brighter than the area around it and much smaller than the size of the real bulb. The image may be only about the size of your fingernail or smaller.

The curved shape of the magnifying lens causes light rays to bend and focus on an image. When we look through the lens, we can use it to make writing or some other objects appear larger. However, the magnifying lens can also be used to make something smaller. The light from the bulb is bent and focused on the wall when the lens is held far from the lamp and close to the wall. The image is much brighter than the surroundings. This is because all the light falling on the surface of the lens is concentrated into a much smaller area.

When sunlight is concentrated by passing it through a lens, the result can be an intensely bright and hot spot of light. Even a small magnifying glass can increase the intensity of the sun enough to set wood and paper on fire. We are using a light bulb rather than sunlight for this experiment because concentrated sunlight can be very harmful to your eyes. NEVER LOOK AT A CONCENTRATED IMAGE OF THE SUN.

The United States Department of Energy's National Renewable Energy Laboratory in Colorado uses solar energy to operate a special furnace. This high-temperature solar furnace uses a lens to concentrate sunlight. A heliostat (a device used to track the motion of the sun across the sky) is used so that the image reflected from a mirror is always directed at the same spot. The lens is used to concentrate sunlight from a mirror to an area about the size of a penny. This concentrated sunlight has the energy of 20,000 suns shining in one spot.

In less than half a second, the temperature can be raised to 1,720° C (3,128° F) which is hot enough to melt sand. This high-temperature solar furnace is being used to harden steel and to make ceramic materials that must be heated to extremely high temperatures.

Concentrated sunlight also has been used to purify polluted ground water. The ultraviolet radiation in sunlight can break down organic pollutants into carbon dioxide, water, and harmless chlorine ions. This procedure has been successfully carried out at the Lawrence Livermore Laboratory in California. In the laboratory, up to 100,000 gallons of contaminated water could be treated in one day.

Other Things to Try

Trace the exact size and shape of the magnifying lens on a piece of paper. Cut out this piece of paper and tape in on the wall. Focus the image of the lamp on this piece of paper and copy the bulb image on the paper. Compare the size of the bulb image to the size of the piece of paper. How much bigger is the lens than the focused image of the bulb? Use this ratio of sizes to estimate the increase in the brightness of the image.

Can you explain why the image of the bulb is upside down when it is projected on the wall? See if you can find information about optics in a book or encyclopedia that could help you explain this reversal of the image.

Repeat this experiment using two magnifying lenses. Observe the effect of moving the positions of the two lenses relative to each other and the wall.

Experiment #13

Can Electricity Be Made From Sunlight?

Materials

Silicon solar cell

Earphone or headset for a portable radio

AA-size battery

Two wires with alligator clips on each end of the wires

Procedure

For this experiment you will need a silicon solar cell. Small silicon solar cells are inexpensive and can be purchased at many electronic supply stores.

This experiment should be done on a bright, sunny day.

Examine the metal shaft on the part of the earphone or headset that is inserted into a portable radio. You will notice that just above the tip of the shaft there is a plastic spacer. Clip on one of the wires below this spacer. Then clip on the other wire above this spacer.

To test that the wires are properly connected to the earphone or headset, take the unconnected ends of the two wires and touch them to an AA-size battery. One wire should touch the positive end of the battery, while the other should touch the negative end of the battery. Place the earphone or headset to your ear. If your connections are made correctly, you should hear a crackling sound in the earphone or headset. If you do not hear a crackling sound, check your connections carefully.

Take the earphone or headset, with wires attached, and the solar cell outside into the sunshine. Ask a friend to join you. Your friend can help you hold the solar cell.

Place the earphone or headset to your ear. Ask your friend to hold one of the flat sides of the solar cell facing the sun. The two flat sides of the solar cell are different. In this experiment, you will determine which flat side must face the sun for the cell to generate electricity.

While your friend holds one of the flat sides of the solar cell facing the sun, you hold one of the alligator clips on the side of the cell facing the sun. At the same time touch the other alligator clip to the opposite side of the cell. As you hold the alligator clips to the cell, avoid blocking the sunlight striking the solar cell.

Ask your friend to turn the solar cell over so that the side that was not facing the sun before now does. Touch a clip to the two sides of the solar cell.

After determining which side of the solar cell needs to face the sun to make a crackling sound in your earphone or headset, ask your friend to hold that side toward the sun. Touch the two alligator clips to each side of the solar cell. Move the alligator clip touching the bottom of the solar cell around the bottom side to keep making the crackling sound in your earphone or headset. Next block the sunlight striking the solar cell.

Observations

Describe the difference between the two sides of the solar cell. Which side must be facing the sun to cause crackling in the earphone or headset when you touch the clips to the solar cell? What happens to the crackling sound when you block the sunlight from striking the solar cell?

Discussion

When you examine your silicon solar cell, you will notice that the two flat sides of the cell are different. One

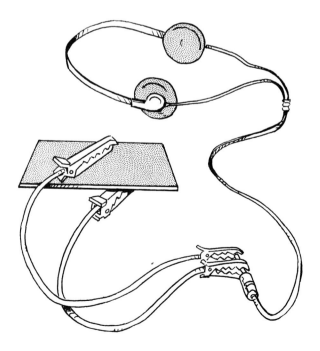

side should have a silvery color, while the other side should appear dark. You should determine in this experiment that one side of the solar cell needs to face the sun for you to hear a strong crackling sound in the earphone or headset. The crackling sound is electricity, generated by the solar cell, passing through the earphone or headset.

The solar cell you are using for this experiment is made from the element silicon. Silicon solar cells consist of two thin wafers of treated silicon that are sandwiched together. The treated silicon is made by first melting extremely pure silicon in a special furnace. Tiny amounts of other elements are added which produce either a small positive or negative electrical charge. Usually boron is added to produce a positive charge and phosphorus is added to produce a negative charge. The addition of these other elements to pure silicon to produce an electrical charge is called doping.

After being doped, the molten silicon is allowed to cool. As it cools, the doped silicon grows into a large crystal from which very thin wafers are cut. A wafer cut from a large crystal of silicon doped with boron is called the positive or P-layer because it has a positive charge. A wafer cut from a large crystal of silicon doped with phosphorous is called the negative or N-layer.

To make a solar cell, a positive wafer (P-layer) and a negative wafer (N-layer) are sandwiched together. This causes the P-layer to develop a slight positive charge, and the N-layer to develop a slight negative charge. The solar cell is connected to a circuit by wires leading from the P-layer and the N-layer. When light falls on the surface of the cell, electrons are made to move from one layer to the other. Thus, a current of electricity flows through the circuit.

The first solar cells provided electrical power for space satellites and vehicles. Satellites and space vehicles are still big users of solar cells. Solar cells are now being used to provide electrical power for calculators and similar devices, weather stations in remote areas, oil-drilling platforms, and remote communication relay stations.

The best silicon cells convert only a small portion of the sunlight striking the cells into electricity. The efficiency of solar cells is about 15 percent. This means that 15 percent of the sunlight that strikes the cell is converted into electrical energy. The sunlight that is not converted into electricity either reflects off the surface of the cell or is converted into heat energy.

Other Things to Try

Repeat this experiment on a cloudy day. Do you still hear a crackling sound in the earphone or headset? If you do hear a crackling sound, is it quieter than on a bright, sunny day?

Is there enough light in a room to cause the solar cell to make electricity? Try it yourself and see.

V. Energy From Biomass

Fossil fuels, which include petroleum, natural gas, and coal, supply nearly 90 percent of the energy needs of the United States and other industrialized nations. Because of their high demand, these nonrenewable energy resources are rapidly being consumed. Some estimates suggest there is only a 500-year supply of oil and natural gas remaining on Earth. Coal supplies are expected to last about a thousand years.

We must find other sources of energy to meet the increasing fuel demands of modern society. Important alternate sources of energy include: solar, wind, biomass, hydroelectric, geothermal, nuclear, and tidal energy.

One of the benefits of using alternate sources of energy is that many of them are "clean." This means that they do not cause pollution. Also, many alternative energy sources are renewable energy sources. They are replaced naturally—such as plant life—or are readily available—such as the sun and wind. In addition, the use of renewable forms of energy will allow us to stretch out our current supply of fossil fuels so they will last longer.

In this chapter you will learn how biomass, or organic matter, can be an important energy source.

Plants are the most important biomass energy source. Plant material can be burned directly—as with wood—or it can be converted into a fuel by other means. In the experiments that follow you will explore: how water can be heated by composting grass, how a peanut burns, and how corn syrup can be made into ethyl alcohol.

Experiment #14

Can Water Be Heated With Composting Plant Material?

Materials

Freshly cut grass clippings

Empty two-liter plastic drink bottle

Water

Cooking thermometer (optional)

Rake

Styrofoam cup

Procedure

Ask an adult to help you collect freshly cut grass clippings. You will need enough grass clippings to fill about two large grocery bags. Rake the grass clippings into a pile on a shady spot in the yard. In some parts of the country you will only be able to do this in the spring and summer months when grass grows.

Fill a two-liter plastic bottle with cold water. Starting at the top of the pile dig a hole down into the grass clippings just large enough for the plastic bottle. Place the plastic bottle in the hole, and then fill around it with grass clippings. The top of the plastic bottle should stick out of the pile of grass clippings slightly.

Check the water in the bottle after it has been in the

pile of grass clippings for twenty-four hours. If you have a cooking thermometer, place it in the bottle for thirty seconds and then read the temperature on the thermometer. If you do not have a cooking thermometer, carefully remove the bottle of water from the pile of grass clippings. Feel the sides of the plastic bottle. CAUTION—THE BOTTLE MAY BE VERY WARM, SO AVOID BURNING YOURSELF. Pour some of the water into a Styrofoam cup and look for water vapor rising from the cup. Place the bottle back into the pile of grass clippings and check it again each day for several days.

Observations

After twenty-four hours in the grass clippings, is the water in the plastic bottle warm when you check it? If

you check the water with a thermometer, what is the temperature of the water?

When you pour some of the water into a Styrofoam cup, do you see water vapor rising from the cup?

How many days does the water remain warm?

Does the pile of grass clippings become smaller after a few days? What does the pile of grass clippings look like when the water is no longer warm?

Discussion

The water in the plastic bottle should be warm after the bottle has been in the pile of grass clippings for twenty-four hours. The heat that warms the water comes from the pile of grass clippings, which is decomposing or composting.

The decomposition of dead plant and animal material is nature's way of recycling important chemical substances. Complex chemical substances in dead plant and animal material are broken down into simple chemical substances during the process. These simpler chemical substances then become nutrients for living plants and soil animals.

Heat is given off during the decomposition process. The more material that is decomposing, the more heat is produced. In this experiment, a large amount of heat is given off by the decomposing grass clippings. This is because you started with a large pile of grass clippings, and grass clippings decompose quickly.

A pile of decomposing grass clippings can reach a temperature of over 71°C (160°F). The water in the bottle may absorb enough heat from the decomposing grass to reach a temperature as high as 60°C (140°F) after a day or two. For comparison, most household hot water heaters are set to deliver hot water with a temperature between 54°C (130°F) and 71°C (160°F).

Other Things to Try

How much water can you heat with composting grass clippings? To find out, repeat this experiment with a one-gallon plastic milk jug filled with water. Does the water in the jug become warm?

Repeat this experiment, but remove the bottle of water from the pile of grass clippings after twenty-four hours. Fill a second two-liter plastic bottle with cold water from a sink faucet and place it into the pile of grass clippings. Does the water in this bottle become warm after a couple of hours?

Is there a minimum amount of grass clippings that are needed to make enough heat to heat the water in a two-liter plastic bottle? To find out, surround a two-liter plastic bottle filled with water with just enough grass clippings to cover it. Does the water become warm after twenty-four hours?

Experiment #15

Do Plants Store Energy?

Materials

Shelled peanut

Small pair of pliers

Match or lighter

Sink

Procedure

ASK AN ADULT TO HELP YOU WITH THIS EXPERIMENT. DO NOT USE A MATCH OR LIGHTER BY YOURSELF.

Close the drain in the kitchen sink. Fill the sink with water until the bottom of the sink is just covered.

Using a small pair of pliers, hold the peanut over the sink containing water. Ask an adult to hold the flame of a lit match or lighter directly under the peanut. When the peanut starts to burn, the lit match or lighter can be removed.

Allow the peanut to burn for one minute. MAKE SURE AN ADULT REMAINS PRESENT AND MAKE SURE TO HOLD THE PEANUT OVER THE SINK. To extinguish the burning peanut, drop it into the water. After you have extinguished the peanut, allow it to cool and then examine it carefully.

Observations

How long does it take for the peanut to start to burn? Does the peanut burn with a clean flame or a sooty flame? What color is the flame? What color does the peanut turn when it burns? Did the size of the peanut change after it has burned for several minutes?

Discussion

You should find that the peanut ignites and burns after a lit match or lighter is held under it for a few seconds.

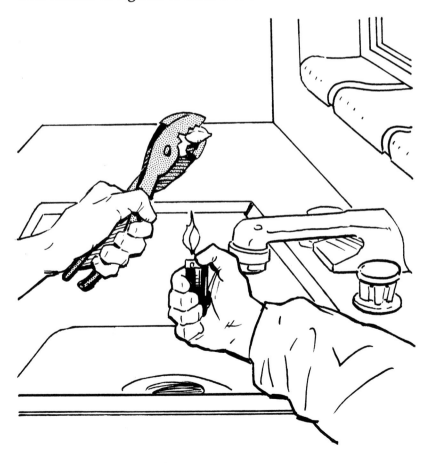

Although you only let the peanut burn for one minute as a safety measure, the peanut would burn for many minutes.

A peanut is not a nut, but actually a seed. In addition to containing protein, a peanut is rich in fats and carbohydrates. Fats and carbohydrates are the major sources of energy for plants and animals.

In this experiment, when the peanut burns, the stored energy in the fats and carbohydrates of the peanut is released as heat and light. When you eat peanuts, the stored energy in the fats and carbohydrates of the peanut is used to fuel your body.

The energy contained in the peanut actually came from the sun. Green plants absorb solar energy and use it in photosynthesis. During photosynthesis, carbon dioxide and water are combined to make glucose. Glucose is a simple sugar that is a type of carbohydrate. Oxygen gas is also made during photosynthesis.

The glucose made during photosynthesis is used by plants to make other important chemical substances needed for living and growing. Some of the chemical substances made from glucose include fats, carbohydrates (such as various sugars, starch, and cellulose), and proteins.

Photosynthesis is the way in which green plants make their food, and ultimately, all the food available on earth. All animals and nongreen plants (such as fungi and bacteria) depend on the stored energy of green plants to live. Photosynthesis is the most important way animals obtain energy from the sun.

Oil squeezed from nuts and seeds is a potential source of fuel. In some parts of the world, oil squeezed from seeds—particularly sunflower seeds—is burned as a motor fuel in some farm equipment. In the United States, some people have modified diesel cars and trucks to run on vegetable oils.

Fuels from vegetable oils are particularly attractive because, unlike fossil fuels, these fuels are renewable. They come from plants that can be grown in a reasonable amount of time. Fossil fuels are nonrenewable fuels because they are formed over millions of years.

Other Things to Try

Hold one end of a piece of uncooked spaghetti in a pair of pliers. Ask an adult to hold the flame of a lit match or lighter under the other end of the spaghetti. When the spaghetti starts to burn, place it in an aluminum pie pan that is in the sink. Make sure to extinguish the burning spaghetti with water when you are finished with the experiment. How does the burning of the spaghetti compare with the burning of the peanut?

Experiment #16

Can a Fuel Be Made From Plant Material?

Materials

One package of yeast

Water

Corn syrup

Empty, two-liter, plastic drink bottle

Balloon

Measuring spoons

Measuring cup

Funnel (optional)

Sink or bucket

Procedure

Remove the paper label from around an empty, two-liter, plastic drink bottle. Add two cups of water and one package of yeast to the bottle. Swirl the bottle to mix the water and yeast.

Next, add one-quarter cup of corn syrup to the bottle. You may want to use a small funnel to help you add the corn syrup to the bottle. Swirl the bottle to mix the contents.

Place a deflated balloon over the neck of the bottle. Make sure the balloon fits securely over the neck. Place the bottle in a sink or bucket. Check the bottle and balloon after two hours, and then again after four hours.

Finally, check the bottle and balloon after twenty-four hours.

When you have finished with the experiment, pour the contents of the bottle down the sink. Then rinse the bottle and sink with water.

Observations

What color is the yeast mixture? Does the balloon start to inflate after an hour or two? Can you see gas bubbles

rising to the surface of the yeast mixture? Does the balloon grow larger with time?

Discussion

You should find that soon after you mix together the yeast, water, and corn syrup, changes start to take place in the bottle. You should notice that a foam forms on top of the liquid mixture. You should also see tiny gas bubbles rising to the surface of the liquid. Also, you should notice that the balloon begins to inflate and become large.

Yeast is a simple living organism that can break down sugars into ethyl alcohol (ethanol) and carbon dioxide. The process by which yeast breaks down sugars into ethyl alcohol and carbon dioxide is called fermentation.

The tiny gas bubbles rising in the liquid mixture in the bottle are carbon dioxide gas bubbles that are made during the fermentation. The balloon on the bottle expands and becomes inflated because it traps the carbon dioxide gas being produced.

The ethyl alcohol that is made during fermentation stays in the liquid mixture. When fermentation is finished, the liquid mixture usually contains about 13 percent ethyl alcohol. The rest of the liquid is mostly water.

The ethyl alcohol can be concentrated by a process called distillation. During distillation, the liquid fermentation mixture is heated to change the ethyl alcohol and some of the water into a vapor. The vapor is then cooled to change it back into a liquid. This distilled liquid contains 95 percent ethyl alcohol and 5 percent water. The remaining water can be removed by special distillation methods to give pure ethyl alcohol.

In some areas of the United States, ethyl alcohol is blended with gasoline to make a motor fuel known as gasohol. About 8 percent of the gasoline sold in the United States is gasohol.

Gasohol burns more cleanly than pure gasoline. This results in fewer pollutants being released into the air. The use of gasohol as a motor fuel is particularly important in cities that have a lot of smog.

Corn syrup is a mixture of simple and complex sugars and water. It is made by breaking down the starch in corn into sugars. The process is called digestion. In this experiment you changed the sugars in corn syrup using yeast. Much of the ethyl alcohol used to prepare gasohol is made by fermenting corn and corn sugar.

Over one billion gallons of ethyl alcohol are made each year by fermentation of sugars from grains such as corn. Ethyl alcohol is a renewable energy source when it is made by fermenting grains such as corn. This is because the grains, such as corn, are easily grown.

Other Things to Try

Repeat this experiment using one tablespoon of table sugar instead of corn syrup. You should find that yeast can also ferment table sugar into ethyl alcohol and carbon dioxide. Table sugar is made from sugar cane and sugar beets. Since sugar cane and sugar beets are renewable plants, ethyl alcohol made from fermenting sugar from these plant products is another renewable energy source.

Complete List of Materials Used in These Experiments

B
bags, zippper-
    close plastic
balloon
batteries,
    AA-size
bottle, plastic
    two-liter
bowl, plastic

C
clay, modeling
corn syrup
cups, measuring
cups, paper
cups, plastic
cup, Styrofoam

E
earphone or
    headset
Epsom salt

F
fan, electric
flowerpot,
    unglazed
    clay
foil, aluminum
funnel

G
grass clippings,
    freshly cut

H
hole punch

I
ice
ice chest or cooler,
    insulated

J
jars, glass

L
lamp with
    incandescent
    bulb
Lysol® spray

M
magnifying lens
match or lighter
mitt, insulated

N
newspaper

O
oven

P
paint, black
paintbrush
pan, plastic
paper clips
paper, white
peanut, shelled
penny
pie pan,
    aluminum
pinwheel
plate
pliers

R
rake
refrigerator/
    freezer

S
saucer, unglazed
    clay
shoebox
silicon solar cell
sink
soft drink can,
    aluminum
spray paint, black
spoons
spoons, measuring
straw, plastic
string

T
tape
tea bags
thermometer
tomato juice
tree, shady

W
watch or clock
water
wires with
    alligator clips

Y
yeast

Index